CAREERS IN SCIENCE

What Do ASTRONOMERS Do?

BENJAMIN PROUDFIT

PowerKiDS press

NEW YORK

Published in 2022 by The Rosen Publishing Group, Inc.
29 East 21st Street, New York, NY 10010

First Edition

Portions of this work were originally authored by Nicole Shea and published as *Be an Astronomer!* All new material in this edition was authored by Benjamin Proudfit.

Editor: Greg Roza
Book Design: Michael Flynn

Photo Credits: Cover abriendomundo/Shutterstock.com; (background texture) Anna Timoshenko/Shutterstock.com; p. 5 Tony Rowell/Corbis Documentary/Getty Images; p. 7 (constellations) Alexandr Yurtchenko/iStock/Getty Images; p. 7 (Galileo) Hulton Archive/Stringer/Getty Images; p. 9 Mediterranean/E+/Getty Images; p. 11 Chakarin Wattanamongkol/Moment/Getty Images; p. 12 Alex Mit/Shutterstock.com; p. 13 Mark Garlick/Science Photo Library/Getty Images; p. 14 Pr3t3nd3r/iStock/Getty Images; p. 15 Jamie Cooper/SSPL/Getty Images; p. 17 Univeral Historical Archive/Getty Images; p. 18 Science & Society Picture Library/SSPL/Getty Images; p. 19 courtesy of NASA; p. 21 Alex Wong/Getty Images; p. 22 Bill Florence/Shutterstock/com; p. 23 Image Source/Getty Images; p. 25 Stocktrek Images/Getty Images; p. 27 André Farinha/500px/Getty Images; p. 28 Hill Street Studios/DigitalVision/Getty Images; p. 29 David Malan/Stone/Getty Images.

Library of Congress Cataloging-in-Publication Data

Names: Proudfit, Benjamin, author.
Title: What do astronomers do? / Benjamin Proudfit.
Description: New York : PowerKids Press, [2022] | Series: Careers in science | Includes bibliographical references and index.
Identifiers: LCCN 2020040290 | ISBN 9781725329515 (library binding) | ISBN 9781725329492 (paperback) | ISBN 9781725329508 (6 pack)
Subjects: LCSH: Astronomers--Juvenile literature. | Astronomy--Vocational guidance--Juvenile literature. | Astronomy--Juvenile literature.
Classification: LCC QB51.5 .P768 2022 | DDC 520.23--dc23
LC record available at https://lccn.loc.gov/2020040290

Manufactured in the United States of America

CONTENTS

SPACE SCIENTISTS

Have you ever looked up at the stars and wondered what was really out there beyond Earth? Astronomers do more than wonder! They're scientists in the field of astronomy, or the study of the stars, planets, and space. Sometimes what's in space, such as stars and planets, are called heavenly bodies or celestial bodies. Astronomers **research** these space objects and space itself by using specialized tools and **technology**. They also have lots of knowledge from years of study about stars, moons, and more.

It is an astronomer's job to ask questions about how celestial bodies form and how they will change. They look for answers about how big the universe is and how it began. Astronomers are **curious** scientists that work in an out-of-this-world career!

ASTROPHYSICISTS?

Astronomers and astrophysicists study the same thing—space! In the past, astrophysics was defined as the study of the **physics** of the sun, moon, planets, and stars, including their movement and how they act in space. Today, astronomers may do that too and these scientists are often called either an astronomer or astrophysicist.

Astronomers study far more than just stars and planets as more discoveries about space are made.

THE HISTORY OF ASTRONOMERS

Astronomy is one of the oldest sciences people study. That means being an astronomer is a job with a long history. People have watched how stars looked and moved for thousands of years. They discovered that Polaris, called the North Star, seems to be fixed in the night sky. Travelers could walk toward Polaris and always be sure they were headed north.

In 1610, an Italian scientist named Galileo looked through a new tool for stargazing, the **telescope**. He saw four small bright "stars" moving near Jupiter. He watched their movements over several nights and learned that these objects were really moons that were circling, or orbiting, the planet Jupiter!

Early stargazers saw patterns in the stars that reminded them of animals or characters from stories. A pattern of stars that has been named is called a constellation.

GALILEO

Galileo Galilei worked in math, astronomy, and philosophy in the 1500s and 1600s. His observations of space supported the heliocentric theory, which says the Earth moves around the sun. Before this, many people believed the Earth was the center of our solar system. In addition to Jupiter's moons, Galileo discovered that Saturn has rings!

Galileo Galilei

BECOMING AN ASTRONOMER

Today, astronomers are more than simply stargazers. They are very educated people who spend a long time studying to do their job.

All astronomers have to go to college. They take many classes in math and science, including a lot of chemistry and physics, or the sciences that have to do with understanding how matter and energy affect each other. These classes help them understand how objects move through space, the celestial bodies found there, and where Earth fits into this picture. Students studying to be astronomers also need to know computer programming in order to work with the technology of astronomy.

Following college, astronomers stay in school to earn a master's degree and often a PhD., or doctoral degree, in physics, math, or astronomy.

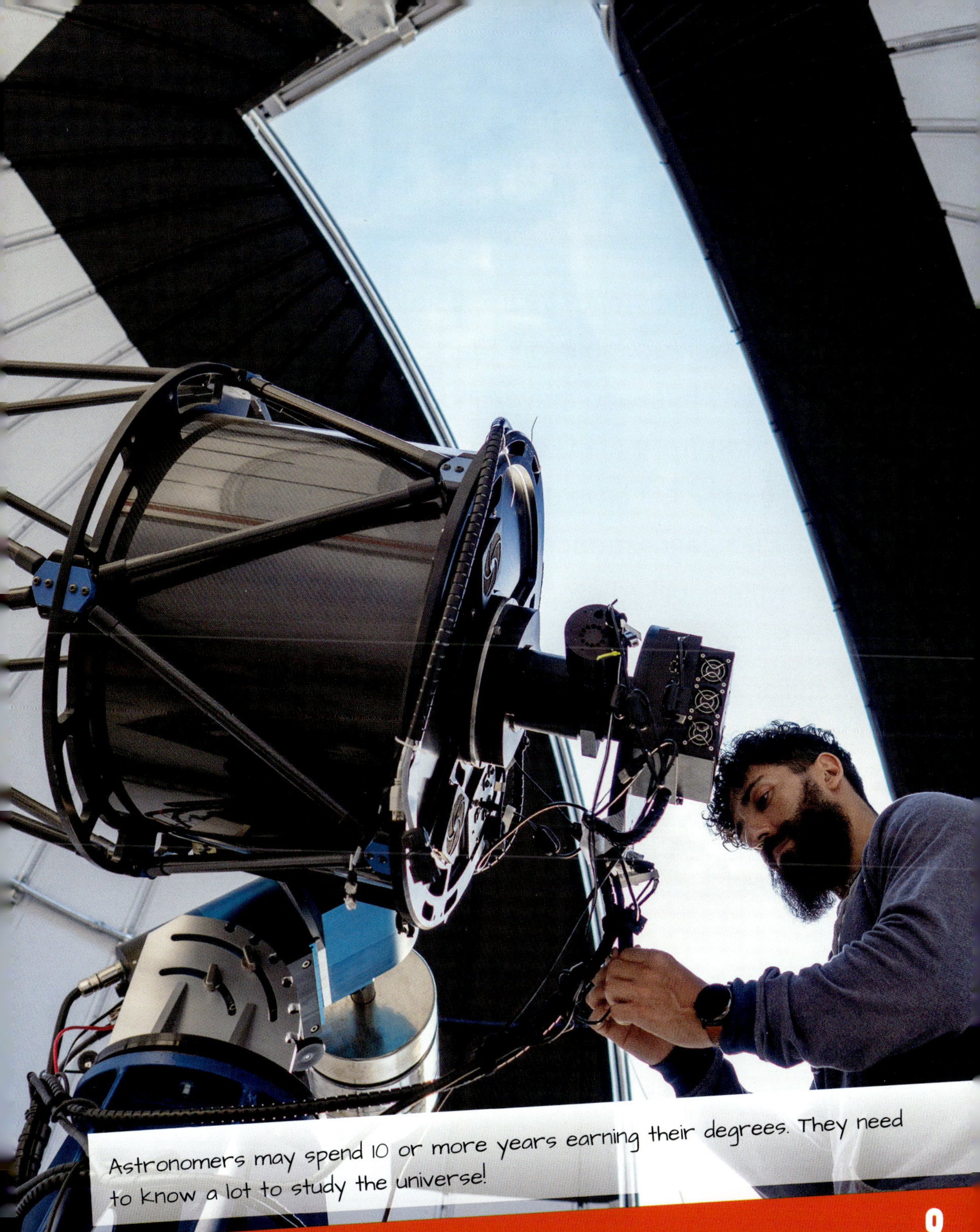

Astronomers may spend 10 or more years earning their degrees. They need to know a lot to study the universe!

STAR SEARCH

During their many years of study, astronomers choose what they want to focus on. Planetary astronomers study how planets form and move, and what their surface, center, and **atmosphere** are like. Thanks to astronomers, we know that other planets have seasons and even storms, just like Earth. Planetary astronomers also study planets beyond our solar system, called exoplanets, and moons.

Some astronomers focus on the study of stars, or stellar astronomy. They research how stars are born, change, and finally die. One of the most interesting things stellar astronomers look for is a huge explosion called a supernova. A supernova occurs when a large star **collapses**. Because the explosion can shine brighter than an entire galaxy, supernovas help astronomers find galaxies they haven't discovered before.

On some nights, planetary astronomers don't need a telescope. Here, the moon, Jupiter, and Venus can all be seen in the sky.

TWO GROUPS

Astronomers fall into two general groups. The first are observational astronomers, including planetary and stellar astronomers. They study stars, planets, and space by observing them closely and learning from these observations. The second are theoretical astronomers. They make models of how stars, planets, and parts of space have changed over time and will change in the future.

Milky Way galaxy

Some stellar astronomers study what no one can see. Black holes form from dying stars. Their mass may collapse until their gravity is so strong that nothing, not even light, can escape. Since there's no light coming from black holes, they're impossible to see. However, astronomers can tell that a black hole is there by tracking its effects on nearby stars and planets. There's a huge black hole at the center of our Milky Way **galaxy**. It's more than 4 million times more massive than our sun!

Today, several telescopes around the world are being used at the same time to observe black holes. When working together, they are called the Event Horizon Telescope (EHT). This international project is putting together data to make pictures of black holes!

A LONG WAIT

The astronomers working with the EHT are proving that black holes exist. They want to know more about black holes and their effects on the space around them. The EHT recorded its first data in 2017. It took a long time for the data to be processed correctly. So, the first pictures took years to put together!

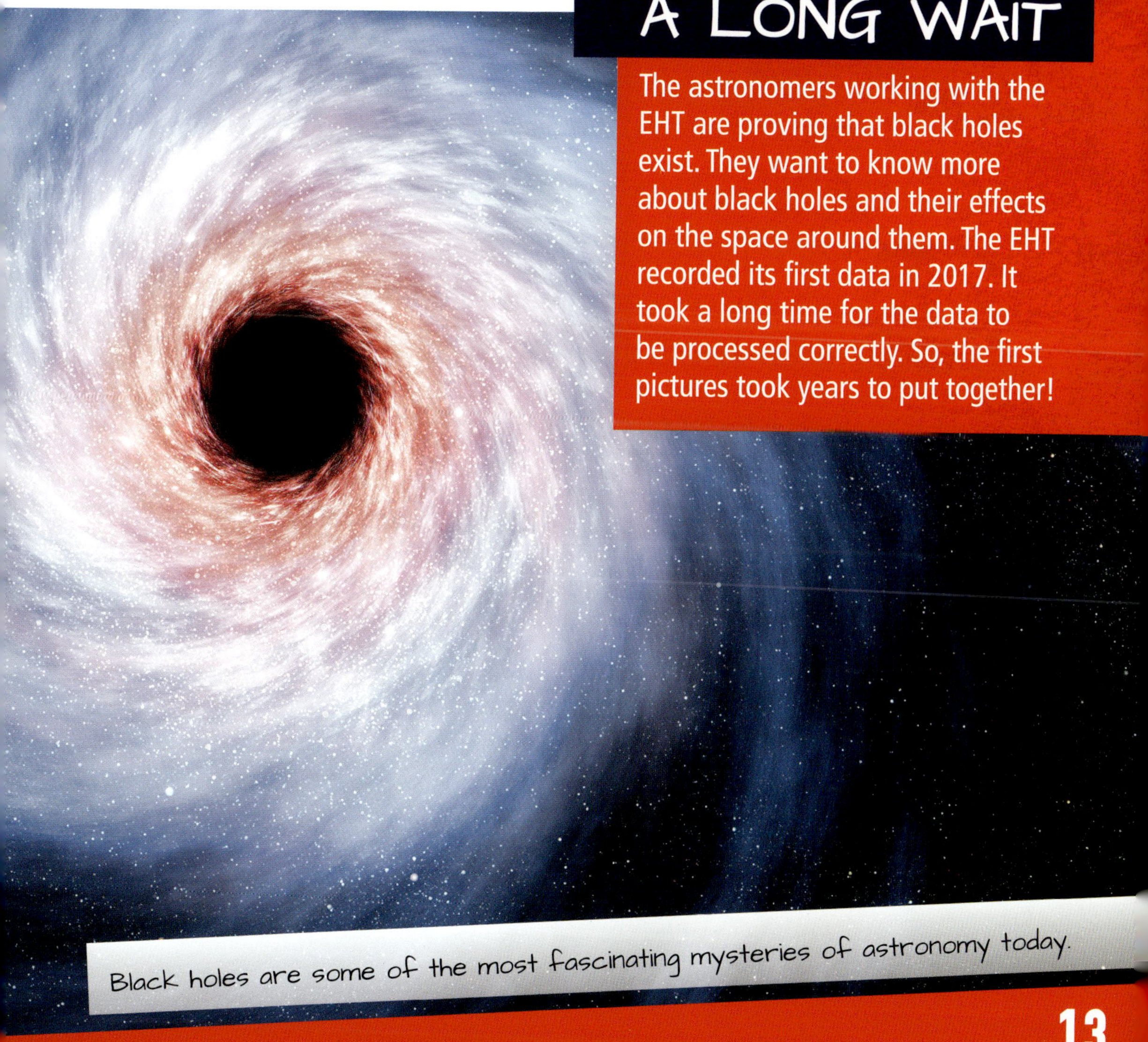

Black holes are some of the most fascinating mysteries of astronomy today.

The most important star in our solar system is the sun. Solar astronomers' job is to study what the sun's atmosphere is like and solar storms, which can affect life on Earth. They may also come up with new ways to study the sun.

Galactic astronomers study galaxies. Galactic astronomers work to learn how galaxies form and change. Many focus on our galaxy, the Milky Way. Extragalactic astronomers find and study other galaxies.

Some astronomers track objects that may come much closer to Earth. Meteors are small burning pieces of rock traveling through Earth's atmosphere. Comets are large pieces of rock, ice, and other matter. Some comets pass by Earth regularly after a certain number of years. Asteroids are made of rock and sometimes metal, mostly nickel and iron.

By comparing our sun to many other stars, astronomers have discovered that our sun is a medium-sized star. Some stars, like red giants, are so big that a few thousand of our sun could fit inside!

OUT OF THIS GALAXY!

There are about 200 billion other stars in the Milky Way galaxy our solar system is found in. Some galaxies are even larger, with more than 100 trillion stars. Astronomers think there may be more than 100 billion galaxies in the universe! That means there are a lot of stars and possible Earthlike planets out there.

THE BIG QUESTIONS

Just as some astronomers study the small objects in space, others may study the universe as a whole. Called cosmologists, these astronomers do research to learn how the universe began and what happened to it in the past. Unlike other astronomers, cosmologists work on many **theories**. This means they don't only study the observable objects in space like a stellar or planetary astronomer might.

Cosmologists' job keeps getting bigger. That's because many scientists think the universe is expanding! Imagine space is like a black rubber sheet with little motorized cars as the stars. Now, imagine stretching the rubber sheet. Not only do the cars (or stars) move, but the space they're traveling on is moving too!

The distance from our sun to the nearest star is about 4.24 light-years. The nearest galaxy is about 25,000 light-years from our solar system. That means that the stars astronomers are looking at appear as they were 25,000 years ago!

LIGHT-YEARS

Astronomers use a unit of measurement called a light-year. A light-year is the distance light travels in one year, which is about 5.88 trillion miles (9.5 trillion km). The farther away something is, the longer it takes for light to travel to us. So, when astronomers see the light from a star four light-years away, they're seeing what it looked like four years ago.

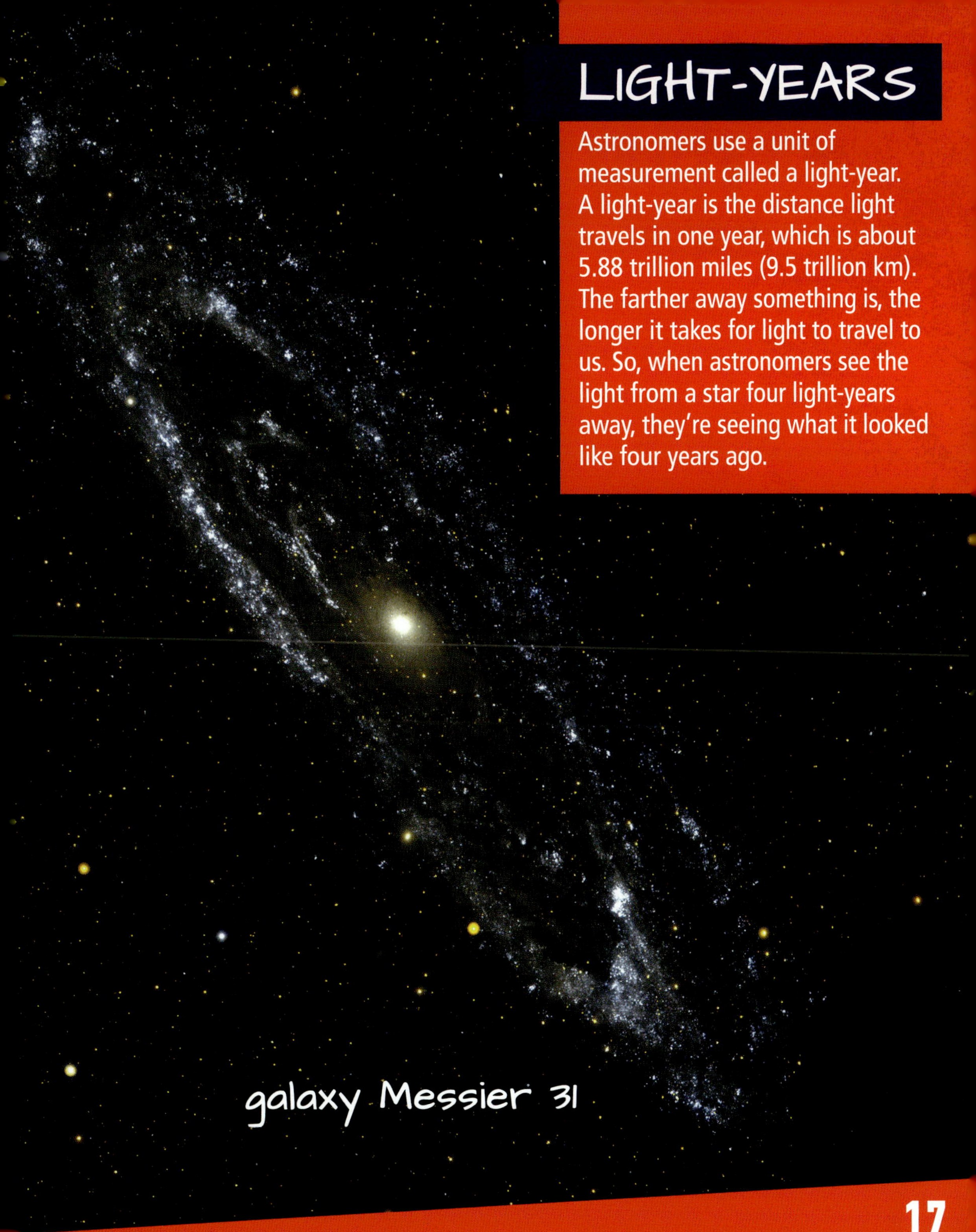

galaxy Messier 31

ON THE JOB

No matter what astronomers choose to study, they do a lot of the same things on the job. They spend time observing space and collecting measurements and other data. This data may be collected by special computer programs. Then, they spend a lot of time **analyzing** this data and data collected by other scientists. Astronomers work with different kinds of scientists to figure out what it means. Sometimes they can only use their knowledge to guess at what it means!

Astronomers may then speak about or publish their findings. It's important to share their analysis, data, and observations so other scientists can use it in future research. It's also a way for more people to learn about space and space objects.

Johannes Kepler

Named for Johannes Kepler, the German scientist who explained how the planets move, the Kepler Space Telescope gathers data many astronomers analyze. Through it, astronomers found a planet that could be like Earth!

Astronomers of all types use many of the same tools to collect their data and do observations. They often use telescopes—but they are far beyond what Galileo used! These telescopes have cameras attached to them. They can review the images captured at any time. Galileo used **optical** telescopes as many astronomers do today. However, these aren't helpful during the day or on cloudy nights. Other telescopes, such as radio telescopes and X-ray telescopes, can be used in daytime and in bad weather. They use waves of energy—radio waves and X-rays—to create images of space objects.

Astronomers also use the data and observations of tools that are in space, such as the Hubble Space Telescope. They can see more about space using these than any telescope based on the ground.

The James Webb Space Telescope is the largest telescope that has ever been sent into space!

JAMES WEBB SPACE TELESCOPE

Since 1990, the Hubble Space Telescope has sent data and pictures to Earth that allowed scientists to understand much more of space and what's in it. The launch of the James Webb Space Telescope is expected to allow for similarly extraordinary discoveries. The JWST is 100 times more powerful than the Hubble!

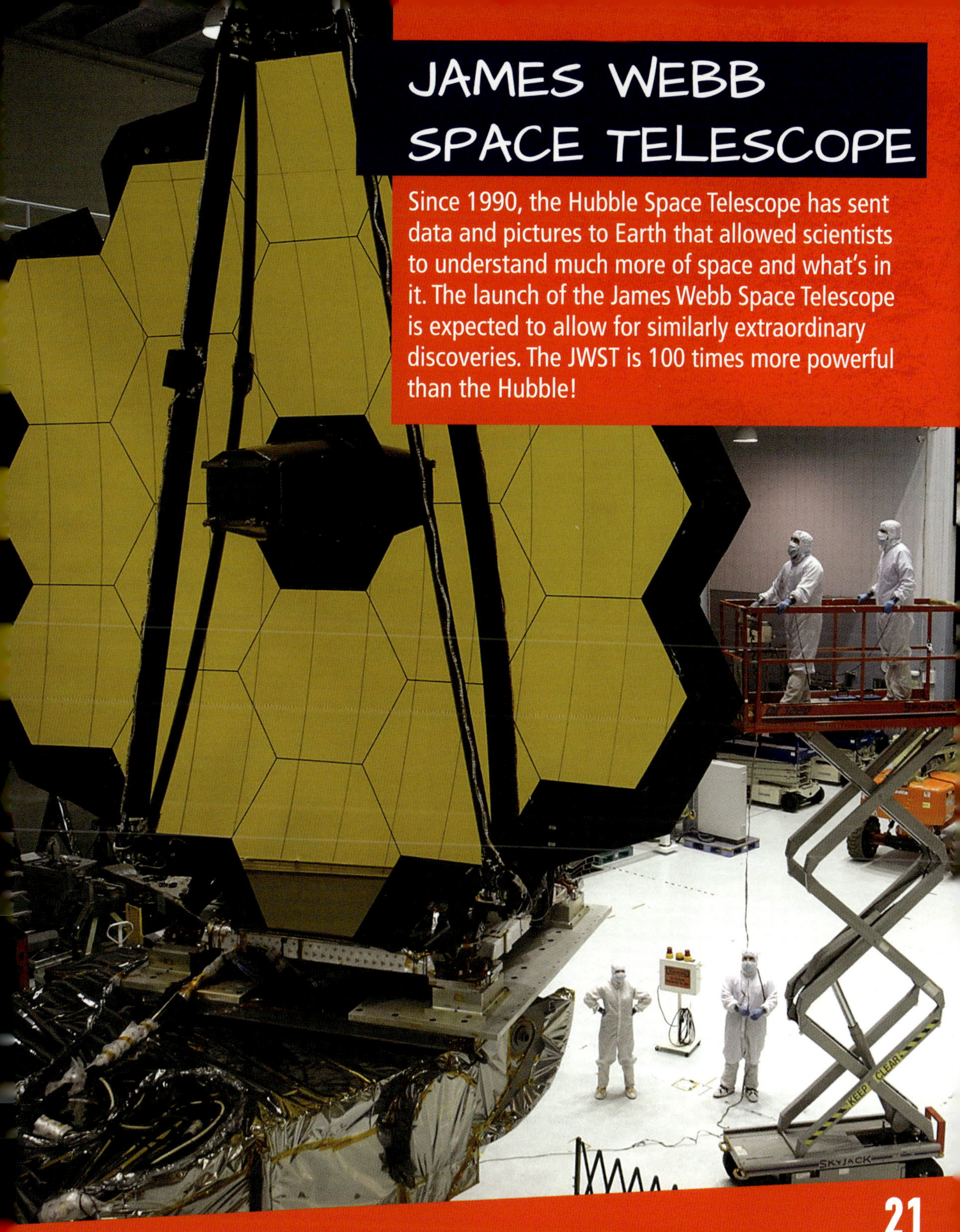

Kitt Peak National Observatory

While many astronomers take part in research as part of their job, it's not always the main part of their job. Many astronomers work at colleges as professors, likely teaching classes in physics as well as astronomy.

Others may work at a ground observatory, such as Kitt Peak National Observatory in Arizona. An observatory is a building specially made for scientists to study the stars, planets, and weather. There, astronomers may have access to some powerful telescopes. Their work might be focused on bettering the telescopes and the research being done at that observatory.

Astronomers may also work at planetariums or science museums. They talk about the solar system and what's beyond it in a way anyone interested could understand.

PLANETARIUMS

A planetarium is a place where you can see what the night sky looks like, no matter the time of day. It has a large room with a curved ceiling and many seats. A special machine shines images on the ceiling and shows the stars and other objects in the night sky. Many planetariums also have telescopes.

Astronomers may work at colleges, planetariums, or observatories around the world. Some travel a lot for work!

Planning and supporting space missions can also be part of an astronomer's job. The National Aeronautics and Space Administration (NASA) uses astronomers to do what they do best: observe space and collect and analyze data. However, the astronomers' analysis might be an important part of launching a probe or even a rocket. NASA is also behind many of the powerful tools astronomers use, such as the Hubble and James Webb space telescopes. They need astronomers' input for building these amazing **instruments**.

Companies working in space and spaceflight may also hire astronomers. These astronomers may do research for special projects. They also may be part of teams that use the company's space tools, such as satellites.

NASA's Cassini spacecraft carried a probe to Saturn. It landed on Saturn's moon Titan and took pictures!

FLYING WITHOUT A PILOT

A space probe is a spacecraft that gathers and sends data back to Earth. A satellite is a machine sent into space to orbit Earth, another planet, the moon, or the sun. Neither has a pilot, or person inside flying it. However, they may be guided by people on Earth—including astronomers!

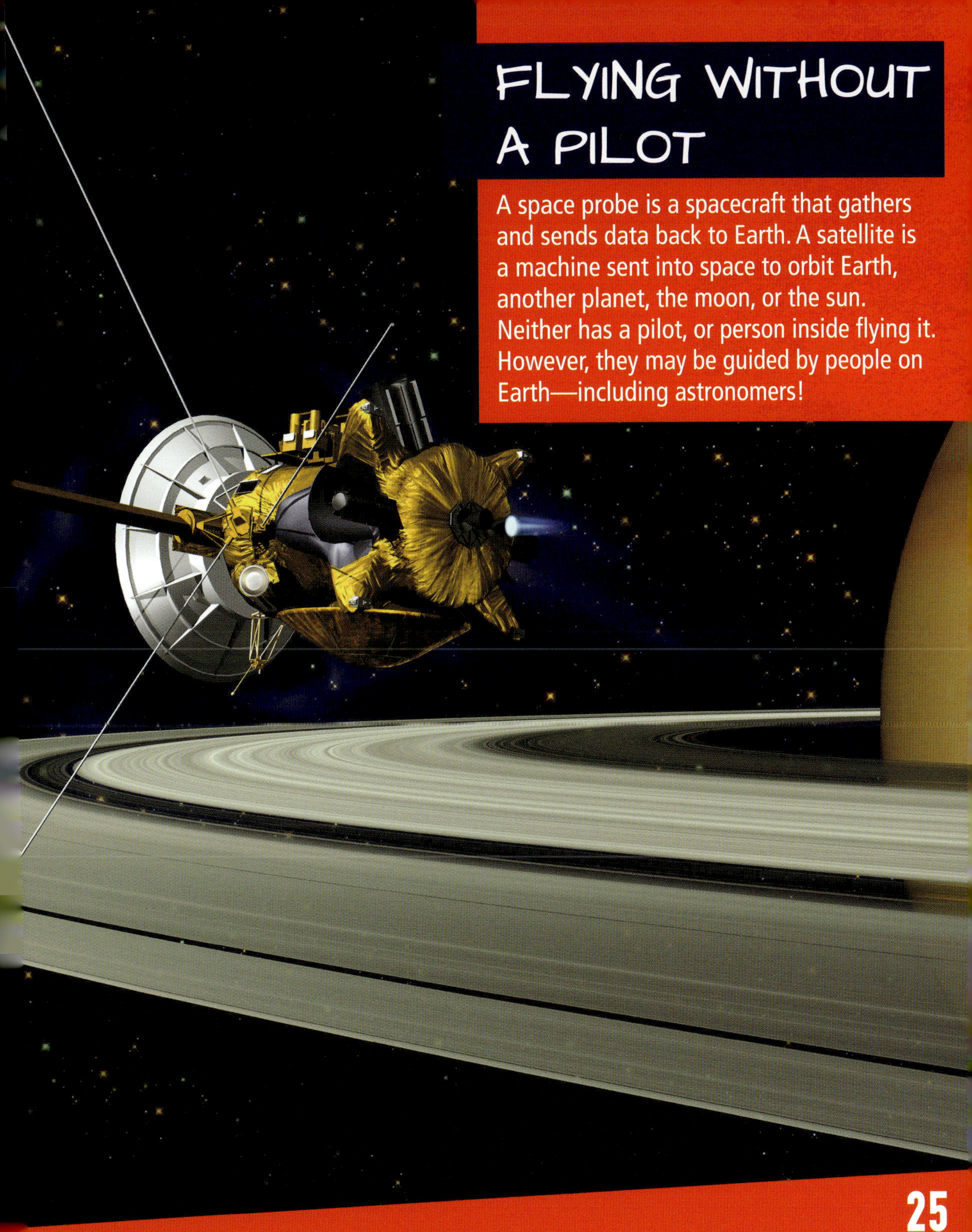

AMATEUR ASTRONOMERS

Becoming a professional astronomer takes many years of school and a lot of patience as they spend time collecting and analyzing data. In fact, there are only a few thousand astronomers in the United States. But not everyone interested in space needs to be a pro at stargazing! **Amateur** astronomers enjoy observing space all over the world. To begin to learn about astronomy, you don't need to sign up for a class or buy special tools.

If you do want to learn more, there are organizations for both amateur and professional astronomers. One is called the Astronomical Society of the Pacific (ASP), and it's for anyone interested in astronomy. The ASP is the largest general astronomy society in the world, with members from more than 70 countries.

If you want to be an amateur astronomer, all you need are your eyes and a clear, starry night! A good telescope doesn't hurt either.

START NOW!

One of the coolest parts of astronomy is that you can start studying this field of science at any age. First, read all you can about astronomy at your school and local library. Scientists are making new discoveries every day, so check NASA's website often to find the most up-to-date information. Visit observatories and planetariums so you can benefit from the latest astronomy technology.

In school, classes in math, science, and computers are all important to astronomy. Astronomers also need to be good researchers and writers. You could even start an astronomy club. Most importantly, be curious about the world around you—and whatever is beyond Earth!

Many cities have observatories or planetariums that welcome school groups. Ask your science teacher if that would make a good field trip!

GLOSSARY

amateur: Someone who does something without pay for enjoyment.

analyze: To study closely and carefully.

atmosphere: The mixture of gases that surround a planet.

collapse: To fall down or cave in.

curious: Having a wish to know and learn more.

galaxy: A large group of stars, planets, gas, and dust that form a unit within the universe.

instrument: A tool used for a certain purpose.

optical: Having to do with light.

physics: The study of matter, energy, force, and motion, and the relationship among them

research: To study to find something new.

technology: A machine, piece of equipment, or method created using science, engineering, and other industries to invent useful tools or to solve problems.

telescope: A tool that makes faraway objects look bigger and closer.

theory: An explanation based on facts that is generally accepted by scientists.

FOR MORE INFORMATION

BOOKS

Barnham, Kay. *The Awesome Night Sky*. New York, NY: Crabtree Publishing Company, 2021

Gregory, Joy. *Astronomer.* New York, NY: AV2 by Weigl, 2020.

WEBSITES

Astronomy: Our Place in Space
www.amnh.org/explore/ology/astronomy
Visit this colorful and entertaining website to learn much more about our place in space.

National Aeronautics and Space Administration
www.nasa.gov
The NASA website is updated frequently with the latest and greatest news about space and astronomy.

Publisher's note to educators and parents: Our editors have carefully reviewed these websites to ensure that they are suitable for students. Many websites change frequently, however, and we cannot guarantee that a site's future contents will continue to meet our high standards of quality and educational value. Be advised that students should be closely supervised whenever they access the Internet.

INDEX